The Girl Who Could Rock the Moon

An Inspirational Tale about Mary G. Ross and the Magic of STEM

Also by Maya Cointreau

The Girl Who Could Write and Unite
The Girls Who Could Fly in the Night
The Girl Who Could Heal Your Heart
The Girl Who Could Sing with the Birds
The Girl Who Could Dance in Outer Space
The Girl Who Could Talk to Computers
Gesturing to God: Mudras for Well-Being
The Comprehensive Vibrational Healing Guide
Shamans Who Work with the Light
The Healing Properties of Flowers
The Practical Reiki Symbol Primer
Simple & Natural Herbal Living
Conversations with Stones
Natural Animal Healing
Equine Herbs & Healing
Grounding & Clearing

The Girl Who Could Rock the Moon

An Inspirational Tale about Mary G. Ross and the Magic of STEM

Maya Cointreau

An Earth Lodge® Publication
www.earthlodgebooks.com
Wallingford,, Vermont

Copyright 2019 by Maya Cointreau
Printed & Published in the United States by Earth Lodge®

ISBN 978-1-944396-87-9

All Artwork, Layout & Design by Maya Cointreau

"I have been lucky to have had so much fun.
It has been an adventure all the way."

Mary Golda Ross

For Eleanor and Sandra,
engineers of today who continue to lead and inspire.
Keep shining!

In the hills of the Ozarks

a girl stared at the stars,

dreaming of travel

to worlds near and far.

Mary "Gold" Ross loved

math, space and science,

so her family found schooling

that would give good guidance.

Talequah city schools

were more than O.K.

Girls and boys learned as equals,

the Cherokee way.

Mary's numbers were magic.

Her facts were pure gold.

She mined figures through college

facing challenges bold.

She earned her degree,

started teaching kids math,

encouraging students

to follow her path.

STEMs aren't just green things

for holding up flowers.

They're fields where minds

play and create for hours.

The S in STEM

stands strong for science.

It brings reason and exploring

into balanced alliance.

T is something

you use at home.

Technology is what

rings your telephone.

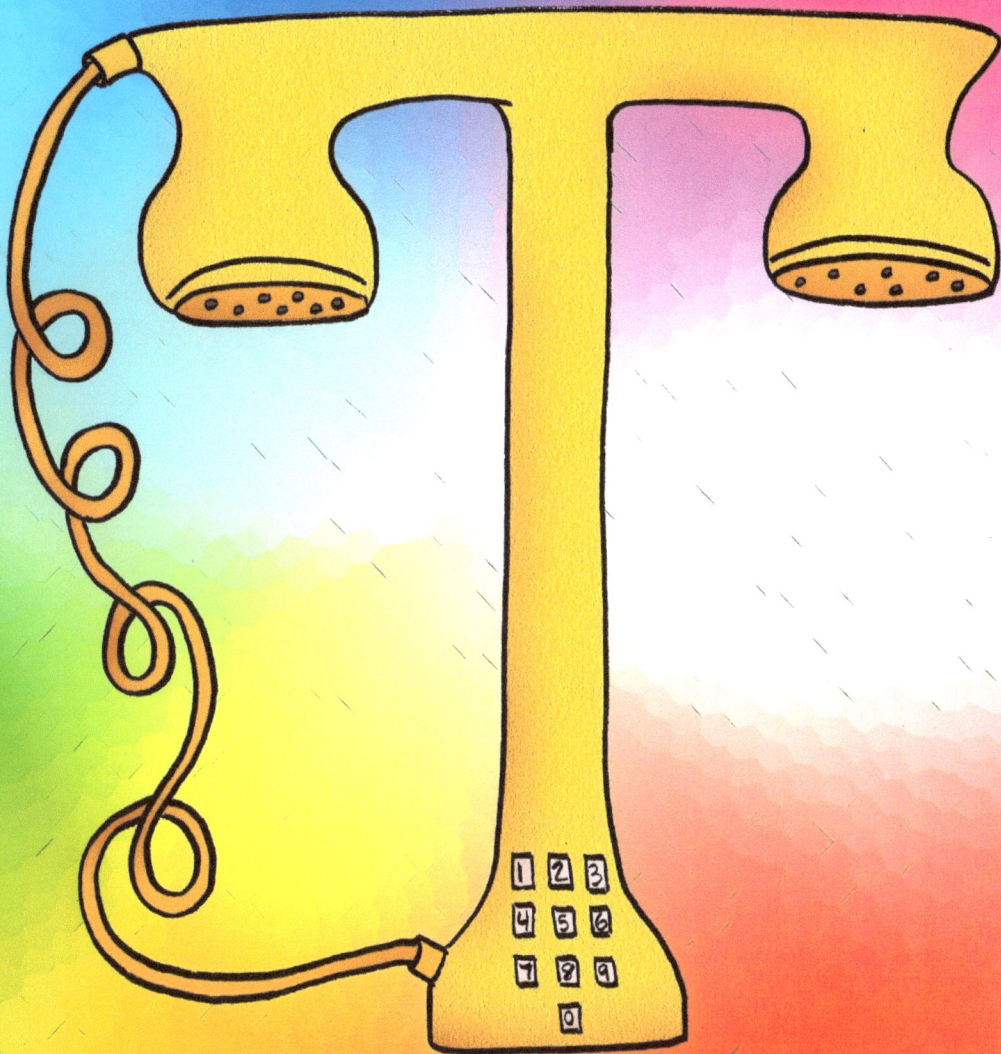

When you get in a car

and watch it go,

you can praise engineering

for making gas flow.

Last but not least,

you have the M.

Mathematics powers

all growth in STEM!

If you want to discover

how good things work

keep practicing math facts

till you get to quarks.

$$K = \frac{1}{2}mv^2$$

$$L = 4\pi d^2 b$$

$$F = \frac{GMm}{r^2}$$

That's what Mary did:

She designed rockets and jets,

using numbers and boldness

to light up sunsets.

When you travel so fast

things quake and break

Mary's math helped NASA

put an end to flight shakes.

Then she wrote a guide

to our solar system,

sent men to the moon,

and she did it with STEM.

So you love to ask questions?

You like technology?

STEM is fun: it's a game.

Go on, start playing like Mary!

More About Mary G. Ross

Mary Golda Ross was born on August 9th, 1908, in Park Hill, Oklahoma. The town was small and she quickly exhibited an intelligence that demanded exceptional schooling. For high school, her parents sent her to live with her grandparents in the Cherokee capitol of Tahlequah. Each day, she walked four miles to get to school, where boys and girls received equal educations. This was unusual for the time, but as a direct descendant of Cherokee chief John Ross, Mary was no stranger to boldness. Ross had led his tribe for almost forty years from 1828–1866, always seeking peace and justice for his people, and was the first indigenous leader to officially petition Congress for tribal grievances. His wife died on the Trail of Tears, a forced tribal relocation over 1200 miles from the Southeastern United States to western states.

Mary had a thirst for knowledge and took advantage of her education. At sixteen, she attended the Northeastern State Teachers' College and received a degree in mathematics at twenty. It was 1928, and the Great Depression had hit the United States. People were out of work, homeless and going hungry. Less than two percent of women were earning college degrees, and even fewer would consider entering STEM fields (STEM is short for science, technology, engineering and mathematics). Mary supported herself in Oklahoma teaching math and science, taking college courses during the summers, and working with the Bureau of Indian Affairs as a statistician.

After ten years, she immersed herself in astronomy and earned a Master's degree in Mathematics from Colorado State Teachers College. The economy had begun to improve, the United States joined the fight against fascism in World War II, and Mary visited California to find more demanding work. She found it at Lockheed in 1942, less than twenty years after Native Americans were granted the right to vote as United States citizens.

At Lockheed, Mary worked as a mathematician solving pressure and flexibility issues for the P-38 Lightning, a military airplane that would fly at speeds exceeding 400mph – one of the fastest of its time. Still, her eye always looked higher. Mary wanted to work on crafts that would fly to space and by the time she retired in 1973, she had: studying satellites and designing vessels that could travel to other planets. Nicknamed "Gold" by her peers, Mary was a valued member at Lockheed. After the war ended, the company sent her to UCLA to study engineering, celestial mechanics, aeronautics and missile design. It wasn't easy going to school in a STEM field when you were a woman. Back then, men dominated the engineering field. But Mary had enjoyed equality during her early years in Tahlequah, and was ready for the challenge.

"I was the only female in my class. I sat on one side of the room and the guys on the other side of the room. I guess they didn't want to associate with me," she said of her time in university. *"But I could hold my own with them and sometimes did better."*

You can watch Mary in 1958 on the famous game show, *What's My Line*, where a panel including actor Jack Lemmon and journalist Dorothy Kilgallen posed questions as they attempted to guess what Mary did for work – finally hitting upon the impressive answer in the end: "Designs Rocket Missiles and Spacecraft."

But Mary wasn't just a designer. She was one of forty elite engineers working with the top-secret Lockheed think tank for Advanced Development Programs, otherwise known as Skunk Works, which lent its efforts to NASA and is credited with helping get man into space and on the moon. Her work on the Agena Space Rocket program helped launch the Apollo Missions. To this day, much of her work remains the foundation of modern space travel – it also remains classified. One of her most famously co-authored publications is volume three of the NASA Planetary Flight Handbook, which details travel protocols to Mars and Venus.

The Agena spacecraft production line at Lockheed
Public Domain Photo Courtesy of the National Reconnaissance Office

Mary never traveled to space herself, but she gave rise to a stronger, bolder generation of women. After she retired, as a long-time member of the Society of Women Engineers, she focused on encouraging young women and Native Americans to enter STEM fields. She supported the National Museum of the American Indian in Washington DC; received awards from the Council of Energy Resources Tribes, American Indian Science and Engineering Society; and was inducted into the Silicon Valley Engineering Council's Hall of Fame in 1992.

In 2002, at the age of 94, she saw John Bennett Herrington of the Chickasaw Nation enter space; he was the first enrolled member of a Native American nation to do so. Two years later, she dressed in traditional Cherokee attire and marched in the opening procession of 25,000 Native peoples at the opening of the National Museum of the American Indian building on the National Mall in Washington, D.C. wearing a traditional green calico Cherokee dress. Said a friend, "She felt she was a part of history being made, again." She died on April 29th, 2008, just shy of her 100th birthday.

Sources: American Indian Magazine, What's My Line, Massive Science, Sacred People Foundation, Purdue Engineering, Unshushed, Oklahoma Today, Lash Publications, USA Today, Newsweek, Wikipedia.

About the Author

Maya Cointreau has been writing and drawing all her life. She lives on a farm in Vermont with her family and thinks nature is the bee's knees. Inspired by her own daughter's interest in puzzles and problem-solving, she began the Girls Who Could series to provide an inspiring template of achievement for all children. Her second book in the series, **The Girl Who Could Dance in Outer Space**, is used as a teaching text in elementary schools nationwide.

For more information about her other books and CDs, visit her website at http://www.mayacointreau.com.

www.ingramcontent.com/pod-product-compliance
Lightning Source LLC
Chambersburg PA
CBHW052048190326
41521CB00002BA/147